Apple TV User's Guide: Streaming Media Manual with Tips & Tricks

By Shelby Johnson
http://techmediasource.com

Contents

Introduction

Deciding on a live streaming option for your entertainment needs may not be as difficult as you think. With just a few viable options on the market, access and reliability play huge parts in which electronic device you choose. Apple TV delivers both in an affordable way, and compact design.

What does Apple TV Do?

The Apple TV media device expands your viewing options and enhances the quality of programming you can receive in your own home. Apple TV, in all of its tiny glory, delivers blockbuster movies, access to your favorite television series, sports programming, music and much, much more by allowing you to stream videos, images and data directly from your iOS device to your television set.

Apple TV vs. Chromecast vs. Roku

Apple TV has two main competitors in the live streaming industry – the Google Chromecast and the Roku streaming media player. Although all three devices allow you to stream movies, television programming and music, Apple hedges both devices by simply having one thing to offer: Better technology. The Roku comes in a number of tiered options, allowing you to choose from a simple version with standard broadcasting capabilities, to one with HD formatting and a microSD card. Apple TV costs the same as the high end Roku version, and provides similar amenities with 1080p HD broadcasting, and the only available version comes remote-ready, with a sleek design. Not all Roku devices include a remote when purchased.

The Chromecast is a heck of a lot cheaper, at a third of the price of Apple TV, but it also has approximately one third of the capabilities, making it an exceptional tool for smaller needs – like streaming from your laptop or tablet directly to the television. It is also a Google product, which means you will need to acquire a Google account if you do not already have one in order to use it.

You can use a Roku, Chromecast and/or Apple TV to save money on monthly cable or satellite TV bills. My guidebook for *How to Get Rid of Cable TV & Save Money* provides additional tips on resources on ditching a monthly cable or satellite television service and using less expensive alternatives to watch TV.

What's in the Box?

As with most things Apple, the design and implementation process for Apple TV is simple. Inside the box you will find:

- Apple TV
- Apple Remote
- Power cord
- Documentation

In order to effectively use the device, there are a few system requirements that must be adhered to in order to stream the content you like best.

What Else Do You Need?

There is a little more to it than just unloading the contents of the box and getting started. You will need the following components to enjoy the benefits of owning an Apple TV, and the good news is each is readily available in this high tech world.

- HDMI Cable (Sold Separately from Apple TV, may be included in certain retailers' package deals)
- High-definition TV with HDMI Compatibility, with a resolution of 720p or 1080p.
- Wi-Fi Wireless Network (or Ethernet cable to plug into your modem)
- iTunes Store Account for Purchasing or Renting Movies, Television Programming and Music
- iCloud Account for Viewing Photos and Videos from your Personal Libraries

In addition, in order to stream movies from Netflix, Hulu Plus or the available sports networks like NHL Center Ice, or NBA.tv, you will need to pay for the subscription to each and enter your username and password for each account to access the content available through Apple TV. It is important to understand that simply because Apple TV comes with Netflix ready does not mean you automatically have access to the content it provides. You will have to subscribe to the service before you can access the content.

Apple TV Inputs & Outputs

While the Apple TV is relatively small and simple as a device, it's important to understand what the various inputs and outputs on the back of the device are used for. Each input or output is described below based on the image below (from left to right).

- Power input – The first input is used for plugging the Apple TV into an available power source.
- HDMI output – Used for plugging an HDMI cable into your Apple TV and outputting hi-definition content to your HDTV set.
- Micro USB – This output is just below the HDMI. It is used for connecting a micro USB cable from your Apple TV to a computer. This is useful for troubleshooting and diagnosing problems with your device. (Apple Care technicians or other professionals generally do advanced troubleshooting and repair using this input/output).
- Optical audio – This is used for plugging an optical audio cable into your Apple TV. The cable will output

high quality digital sound to a stereo receiver or other capable device.

- Ethernet – The final input on the back of the device is used for connecting Apple TV to a home internet network. An Ethernet cable is plugged into your Apple TV, and the other end of the cable is connected to a modem. Ethernet is used in the absence of a wireless or Wi-Fi network.

Initial Setup of Apple TV

To get started with your Apple TV setup, use the following step by step instructions.

Step One: Connecting the Cables
- Choose the exact setup that matches the ports on your television or receiver.
- To connect a widescreen TV with an HDMI port:
- Adapt one end of an HDMI cable to the back of your television.
- Adapt the other end to the back of Apple TV in the proper position.

Step Two: Connect the Power Cord
- Adapt one end of the power cord into the back of Apple TV and the other end into an available power outlet.

Step Three:
- Turn on your television and select the input button, switching to the HDMI connection option.

The first time you use Apple TV you will be met with a series of steps to get started, including choosing a language, selecting a network, configuring Apple TV to work with your network if necessary, and connecting to iTunes.

Configuring Your Apple TV

Have your network password (the same one you use to connect other electronic devices, like laptops and tablets to your WIFI system) ready to go, so you can breeze through the setup effectively. Also, keep your Apple TV Remote at the ready during the configuration process.

If you use a wired Ethernet network to connect, Apple TV will automatically detect your network. If you are using a wireless network to connect, Apple TV will help you access the correct connection that will allow you to configure your network. Simply select your network from the list generated by the device, and enter your network password when prompted.

Set Up Home Sharing

In order to access all of the items from your iTunes library from your Apple TV, you will have to set up home sharing. This is a pertinent step in getting the most from your Apple TV service. First you have to ensure that your iTunes is updated accordingly, as accessing the content in your iTunes library from Apple TV, will require iTunes 7.6 or later installed on your computer.

If you have not updated your iTunes account to at least that version, you will need to do so from your computer before proceeding with the Home Share setup. You can download the latest version of iTunes at http://www.apple.com/itunes/download.

After you set up your network connection, a five-digit passcode will appear on your television screen: Make a note of it!

To set up Apple TV with your iTunes library:

1. Open iTunes on your computer, and update accordingly where applicable.
2. Select the Apple TV icon with "Click to Set Up" directive next to it in the Devices list.
3. Enter the 5-digit passcode from your TV screen.

After you enter the passcode, you can rename your Apple TV and set up iTunes to manage your content on your television screen. If this seems like a ton of work, do not worry. It is much simpler than it sounds and can be accomplished in minutes.

Watch Apple TV

Watching Apple TV happens effortlessly once the setup is completed. You will see all of the options available to you, beginning with new movies and blockbuster releases at the top of your screen, while having access to the seemingly innumerable content listed below it, like television programming, sports, news and family entertainment. In order to navigate these options, you must first familiarize yourself with the remote.

Using the Remote

The simplicity of the Apple TV remote provides intuitive functionality. Simply put, there are only so many buttons to use, so it is incredibly hard to go wrong.

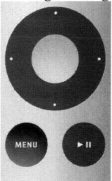

1. To move through menu options or content choices, use the four directional dots on the remote. These allow you to move left, up, right or down on the screen.
2. To select an option from a menu: Press the large silver circle button, or play/pause button.
3. To return to a previous menu: Press Menu.

4. To return to the main menu: Press and Hold Menu Button.
5. To put Apple TV in standby mode: Press and Hold the Play/Pause Button for approximately six seconds.
6. To reset Apple TV: Press and hold the Menu and – (minus) Keys for approximately six seconds.
7. To pair Apple TV and a remote: Press and hold the Menu and Fast Forward Buttons for about six seconds.

It is important to understand that the + and – buttons on some Apple TV remotes DO NOT control your television's volume. You will use the same remote that includes the "input" button that switched your connection to Apple TV to begin with to raise and lower the sound.

Channel Overview

At Apple TV, not all channels are created equal, so it is important to familiarize yourself with the options that are available to you through the streaming device. While there is free programming that is plentiful, you may be required to setup an account with the programming market itself to access their content.

Other subscription services, including Netflix, Hulu Plus and some of the sports channels that are available like MLB.TV, NBA and NHL will require you to pay for their services. If you already have a Netflix or Hulu Plus account and your subscription is current, you can access that programming by simply entering the username and password that is affiliated with each account.

Other channels, including HBOGo or WatchESPN will require you to enter your cable or satellite TV username and password information (or account information, depending on the source of entertainment), assuming you already subscribe to these services from the network providers.

The offerings available through Apple TV are exciting, and a few of the standout channels that are represented within the device are exceptional sources of entertainment.

Netflix

Netflix is available to subscribers, and you account must be up to date to enjoy its content. For $7.99 a month, you can become a member and enjoy the seemingly endless supply of content that is available on demand, whenever you are ready to watch! With your Netflix subscription you will receive complete access to all of your favorite on demand selections, and literally thousands of movies and television shows for your enjoyment.

With Apple TV you can stream movies, television shows, trailers and original directly to your larger screen instead of using your tablet or computer as an entertainment source. With thousands of streaming options available with Netflix, there is no end to the content you can enjoy directly from your Apple TV box. See more info at Netflix.com website.

Hulu Plus

For the same price as Netflix, $7.99, Hulu Plus users can enjoy their favorite movies on demand, and even new network offerings as early as the day after they aired originally. The Hulu database includes over 2000 movies, trailers for new and upcoming movies in the theater, and promos for your favorite shows so you know what to watch for! Check out the Hulu website for more information.

MLB.TV

With MLB.TV, subscribers can watch each and every game throughout the Major League Baseball season, without fail – including a live in-game DVR component, so you can stop, review or pause the game whenever you would like. With multi-game viewing available, you can not only watch baseball through your Apple TV, but can take it with you on the go with your smartphone or tablet. Pricing ranges $19.99 to $24.99 per month, or $109.99 to $129.99 per year.

NBA

Much like the MLB.TV, enjoying the NBA all season long comes with a price. The NBA has two options in this arena, with the first being NBA TV that provides game coverage, news and analysis.

Secondarily, they have the NBA League Pass that is available with a subscription and includes regular season, out of market games for a price of $109 per year. You will have access to each of the games that are on that day or night and can switch back and forth as you see fit. See more info at NBA League Pass.

NHL

As the sporting trend continues, NHL is available through your Apple TV device as well in differing capacities. For $49.95 per season, viewers can watch in game highlights, analysis, live broadcasts, post-game replays and classic games whenever they would like. This gives you access to 40 out of market games per week, so you can follow your favorite team no matter where you live. More info at NHL.com website.

Trailers

Apple TV provides users with access to upcoming movie trailers for films that are currently in the theaters, and those that have yet to debut, giving you a leg up on the entertainment industry and watch to watch for at the cinemas. Likewise, new season previews of your favorite television shows are available, as are trailers for movies coming directly to DVD. This gives you the insight you need to ensure you do not miss your favorite characters by renting the movie they are in, or getting to the theater when it comes out.

YouTube

There is nothing better than navigating the hilarity – and the actual informative content – that is available on YouTube. Instead of playing the website's content on your device, you can literally stream it through Apple TV, giving you all of your favorite videos directly on the big screen. Also, if you have a YouTube.com account, you can access all of your favorites right away simply by logging in.

Vimeo

Vimeo is a video sharing site that allows users to upload, watch and share videos – much like YouTube, but with a focus on longer content. This can include webisodes, shorts and movies that are produced by individuals and shared with the masses. The name is derived from the combination of words "video" and "me."

Podcasts

Podcasts are similar to any other media broadcast, using a host and guests to conduct interviews, analysis, topical and even comedic content that allows individuals to physically stream the material whenever they are ready to listen to the programming – instead of tuning in at a specific time. The listener can subscribe to certain podcasts to have them queued and ready upon their release, so when you turn your Apple TV on, your favorite podcasts will await.

Radio

One of the greatest things about having Apple TV is that you have direct access to your iTunes account from your television, which allows you to not only listen to your collection of favorites, but enjoy iTunes radio! This service is free, but ad supported, and allows listeners to Users are able to skip tracks, customize stations, and purchase the station's songs using their iTunes account, where it will then automatically add the selection into their library.

iCloud Photos

Taking photos on your iPhone or iPad is a great alternative to having a camera at the ready at all times. What's more is that with Apple TV, you can access those images from your television set using the iCloud Photos and share them with whomever is in the room! Even if you take a picture a week ago (or longer!), it will store into the iCloud, and allow you to access it from each of your iOS devices, including the Apple TV.

Flickr

Flickr, pronounced Flicker, is a video and image hosting site that allows account holders to embed their photos or videos for the use of others, creating an online community. Although it can operate as a blog and social media website, it is also a great place for bloggers and social media marketers to find content for use on their pages. The site reports approximately 3.5 million new images being uploaded each day.

WSJ Live

WSJ Live is a mobile version of the Wall Street Journal, which allows individuals to stay up to date with the largest newspaper by circulation in the United States while receiving live alerts, international, national and local updates in real time.

HBO Go

HBO Go is available to HBO subscribers, and in order to access the content on this channel through Apple TV, your cable or satellite carrier information must be made available before access is granted. Simply put, you will have to create an HBO Go account, where you enter your carrier's account number, as well as a created username and password. Once you have done so, you can watch all of your favorite HBO programming on demand, so you never miss an integral episode of your favorite drama again.

PBS

PBS, or the Public Broadcasting System, is the most prominent provider of public television in the United States, boasting over 350 channels nationwide. With access to time-honored programming like Sesame Street, Frontline, Masterpiece, Antiques Road Show – and, of course, Downton Abbey – the channel is consistently named as a most trusted national institution.

WatchESPN

WatchESPN is a great outlet for sports – but only for cable subscribers who already receive ESPN as part of their programming package. If you do, you are in luck and are able to watch all ESPN3 content from your Apple TV, as well as enjoy simulcasts that are currently playing on ESPN, ESPN2, ESPNU, ESPNEWS, ESPN Deportes, ESPN Goal line and ESPN Buzzer Beater, so you will never miss out on your favorite sporting event again. See more details at the WatchESPN FAQ.

WatchABC

Much like the WatchESPN channel – as the networks are owned by the same company – individuals must have a cable network subscription to tap into everything ABC has to offer on demand. Should you, you can watch all of your favorite ABC programming at a moment's notice, including sports, soap operas, sitcoms, comedies and dramas that make the network so popular. View more info at WatchABC website.

Crackle

Crackle is a Sony owned company that provides original web content, feature films and television programming that is produced by Sony Pictures. Much like Hulu Plus or Netflix, users are able to enjoy popular films that exist within the Crackle catalog. The catalog updates monthly, with new content being added and old material being removed to keep the site running tightly.

Bloomberg

Bloomberg television is a 24 hour per day financial news outlet that provides up to date information that affects that financial world, including trading, tickers and informative broadcast analysis. The channel is available by cable or satellite subscription only, and can be viewed from your Apple TV once your account information is entered accordingly.

Smithsonian

Just as you would imagines, the Smithsonian Channel's content is inspired by their research, museums and gives viewers full access to clips, full episodes and content that is available on an array of their networks. The subject matter of the content includes original non-fiction programming that covers a wide range of historical, scientific and cultural subjects. It is only available to viewers who currently subscribe to the channel via cable or satellite enrollment.

Red Bull TV

Red Bull TV provides access to sports, music and live entertainment for free. The channel focuses on extreme sports, and other twists and turns that may not qualify under the same genre, but are nonetheless exciting to watch (including a winch slip and slide "ride"). Surfing, BMX, motorbikes and snowboarding are all available in extreme capacities on the channel, as is some insightful original programming from promoters, athletes and professionals in the sports and music industries.

The Weather Channel

The Weather Channel is delivered directly to your Apple TV device, allowing you to watch real time weather updates in your city or one far away that you are simply curious about or visiting soon. With live updates, forecasts and original programming focused on weather conditions, you will always know what to wear before leaving the house.

Yahoo! Screen

Owned and operated by Yahoo!, Yahoo! Screen is a video sharing website that allows individuals to view editorially-featured videos that change daily and focus on comedy, viral videos, talent, oddities, animation, and premium entertainment content.

Disney Channel

The Disney Channel is available to cable and satellite subscribers and focuses on content for children and pre-adolescents, but is quickly diversifying to include content for teenagers and adults.

Disney XD

Disney XD is a sister channel of the Disney Channel and is aimed at young males, ages 7-14. The content is available through a cable or satellite provider, and provides action and comedy programming to fit its demographic.

Disney Junior

Disney Junior focuses on preschool-age kids, both boy and girl, and delivers access to movies and original programming. It is also only available with a cable or satellite account.

Sky News

Sky News is available without a subscription service on Apple TV, and is a British-based 24 news outlet that focuses on breaking international news. This news outlet is similar to CNN, but does not require a cable or satellite subscription to enjoy.

KORTV

KORTV is available in real time streaming and on demand content, delivering a focus on South Korean movie, television and premium channel content in more than one hundred countries worldwide. The content is available by monthly subscription directly with KORTV, and not a cable or satellite providers – much like a Hulu Plus or Netflix.

iMovie Theater

In an effort to enhance the user's video production from their mobile devices, iMovie Theater was added to Apple TV as a cross device feature designed to allow users to watch shared clips, movies, and trailers on all Apple devices, including the Apple TV. Projects that are created within iMovie on the iOS or Mac platform will automatically show up within the Apple TV channel.

Vevo

Vevo provides direct access to music videos from two of the three major record labels, UMG and SME. This channel provides unlimited access to your favorite artists music videos, reverting back to a time when MTV was paramount programming. The concept of the channel is described as being the "Hulu of music videos" allowing viewers to select videos they want to watch on demand. The service is ad-based, so viewers will have to endure advertisements before watching their choice of videos.

Qello

Qello provides a focused approach to entertainment by streaming only concerts and music documentaries. The channel licenses a variety of long-form concerts, documentaries, behind the scenes footage and interviews from both major and independent music labels. The musical genres are versatile and plenty, ranging from the 1920s to today, with each category covered with in depth original programming.

Crunchyroll

Crunchyroll is an American website, although its content in completely based on East Asian influences. These influences include manga, anime, drama, music, electronic entertainment and auto racing for the enjoyment of an International online community. A great portion of the content is free, and available to the masses, with some programming options requiring an optional membership fee of $6.95 per month.

MLS

Major League Soccer comes to life on Apple TV, but only with a subscription to the service. Subscribers can watch live, out of market games or enjoy interviews with players, and analysis from professional athletes who have seen it all.

WWE Network

The WWE Network is a brand new platform that officially launched for Roku on Monday, February 24, 2014. This service, which costs $9.99 per month as of this publication, allows fans to access a wide library of old and new video content including past pay-per-views, highlight shows and brand new exclusive programs. To access the service, viewers will need to sign up for an account for the WWE Network at WWE.com website, or it can be purchased as a monthly subscription through your iTunes account.
Once you're signed up, you'll need to start up the WWE Network channel on your Apple TV and enter your subscription sign-in details to access the live streaming content on the channel.
See more details about the new 2014 service at: http://www.wwe.com/help.

Apple TV Channels

In addition to all of the significantly branded options like Disney and ESPN – or sporting availability from the professional organizations themselves – there are a number of Apple TV channels that are added randomly, highlighting your love for television.
Celebrating the 50th anniversary of the Beatles, Apple TV created a Beatles channel to provide open access to one of the world's most beloved bands delivering interviews, old footage and new material by the remaining artists.

Although the randomness of the content Apple adds cannot be forecasted, the device does provide event releases and promotional information that allows everyone to stay ahead of the curb, and anticipate the newest channels that will be released. Check your options regularly, and enjoy what's new as it becomes available.

Renting or Buying Media

Apple TV will give you the option to purchase or rent programming, as you see fit. If you would like to stream a movie a single time – renting may be the way to go. If you know that you are going to watch it over and over again, purchasing the movie will allow you to add it to your library to watch as many times as you would like.

The fee associated with renting or purchasing content will be made available on screen, after you select the title, so you can choose your option without issue. The fee will be charged to your iTunes account and the payment method associated with it.

Music, Movies, and TV shows

To access programming from your Apple TV, simply follow these instructions for each category of rental or purchase. Purchasing Television Programming
   1. Select TV Shows from the main Apple TV menu.
   2. Choose Top TV Shows, Genius, Genres, TV Networks, or Search for Programming by Name

3. Find a TV show you would like to purchase and select it. You will have the option to add the show to your Favorites, buy all episodes (a season), or you can purchase individual episodes.

Renting or Purchasing Movie Content
1. Sign in to the iTunes Store. From the main Apple TV menu, choose Settings > iTunes Store.
   Note: If your location and iTunes account already appear here, skip to the next section. If not, Select Location. Scroll to highlight your location and select it. Press the Menu button on the remote once to return to the previous screen.
2. Select Sign in.
3. Enter your Apple ID and password.
   Note: You must have an iTunes account to use these services. If you do not have an Apple ID, create one using iTunes on your computer.
4. Apple TV will give you the option to remember your password. Select Yes if you want to save the password, select no if you want all movie rentals and purchases to be password protected.
5. Press and hold the Menu button to return to the main Apple TV menu.

To Rent (or Purchase)
1. Select Movies from the main Apple TV menu.
2. Choose one of the following categories to begin finding movies:
   • Top Movies
   • Genius
   • Genres
   • Search

3. Find a movie you would like to rent (or purchase) and select it.
4. Click Rent or Purchase.

No matter what you are watching or listening to, you can add restrictions to your rental and purchasing options by setting security measures that will keep others from using your account as they see fit. When setting up your Apple TV, you will be asked if you want to save your password: Click "No." This will create a barrier for anyone who does not have the password from downloaded content that will be charged to your iTunes account.

Hiding Unwanted Channels

As you become familiar with your Apple TV device, there may be certain channels that just do not appeal to you – or that you would rather the rest of your family not have access to. To hide channels, simply follow these instructions – from your Apple TV:
1. Select the Settings Icon
2. Select General
3. Select Parental Controls (or Restrictions, as listed on differing models)

From this section you will be given a list of channels, with the word "Hide" next to it. Click on the channel option to change the text to Hide (or Unhide, if you want to restore its availability).
If you have updated your Apple TV to version 6.1, you can hide channels even easier. One of the new features of the recent Apple TV version 6.1 update is it allows you to hide unwanted channels with ease. If you have updated your software, simply do the following to hide unwanted channels:

1. Choose a channel from the main menu.
2. Hold the Select button until the icons dance.
3. Use directional buttons to move channel icon, and push the Play/Pause button to hide the channel.

Using Music / Computers (to stream from iTunes Library)

If you have an iTunes account, you can play your music from any computer, iPod, iPad or iPhone simply by ensuring your username and password are entered correctly in the settings area. This approach also applies to iCloud, so you can share all of your information – music, movies, images and videos – across all of your iOS platforms.

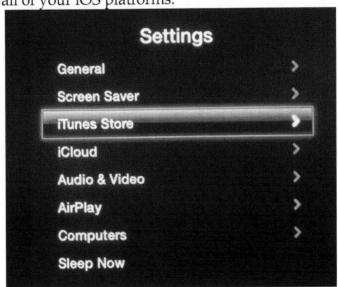

Sync your Apple TV with iTunes

1. Make sure the computer you want to sync with is on, and iTunes is open.
2. On your Apple TV, choose Settings > Computers > Add Shared iTunes Library, and note the passcode displayed.
3. Click the Devices button near the top of the iTunes window to access your Apple TV, and enter the passcode.
4. Click the category buttons (Movies, TV Shows, and so on) and specify your syncing preferences.
5. To transfer a rented movie from your computer to your Apple TV, click the Movies button, select the movie you want to transfer, and click Move.

Now you are all set to enjoy your entertainment from literally anywhere!

iCloud Photos & Photo Streams

Viewing your iCloud photos and photo streams that you share with others on your big screen is incredibly easy, thanks to Apple TV. The first step, of course, is setting up iCloud on your Apple TV, so you can share away! All you have to do is turn on your Apple TV, and enter your Apple username and password to get started. You will then have access to iCloud Photo Sharing and My Photo Stream from your iPhone, iPad, iPod touch, Mac or PC. After signing in with your iCloud account, you can view your photos by choosing iCloud Photos or Photo Stream from the main

What is iCloud Photos & Photo Sharing?

iCloud stores all your photos added by your individual Apple devices into My Photo Stream. This means if you take a picture with your iPhone, it will be stored effortlessly into your iCloud Photos account, and can be managed from any of your other devices that have an Apple account.

You can share these photos and videos by creating a shared stream, and then inviting family and friend to their photos and videos of the same event – whether it is a vacation, reunion or day outing you enjoyed together – from their devices that share the same software. Everyone is free to upload their content and comment on the accessed shared streams as they see fit.

Sharing Photo Streams

You can create a photo stream, beginning with a single photo – or dozens of them, if you see fit. First, you have to turn your photo stream on. Here is how: Make sure your devices are using the latest iOS software. Set up your iCloud account and turn on "My Photo Stream" on any of your devices:
On ANY iOS Device:

- Select: Settings.
- Select: iCloud.
- Select: Photos (Photo Stream in iOS 6).
- Turn On: My Photo Stream.

On ANY Mac Computer:

- Select: System Preferences.
- Select: iCloud.
- Select: The Checkbox for "Photos" or "Photo Stream."
- Select: Options.
- Verify that "My Photo Stream" checkbox is selected, and open the application you want to use and confirm that My Photo Stream is turned on.

On ANY PC:

- Open: iCloud Control Panel for Windows.
- Select: Checkbox for Photos (Photo Stream in iCloud Control Panel 2.x).
- Verify that "My Photo Stream" is enabled and adjust the settings and click "Apply" when finished.

Keep in mind that photos uploaded to "My Photo Stream" do not count against your iCloud storage memory, and that when the application is enabled all new photos you take will be automatically added to your stream. The images will be stored for thirty days, before being downloaded to your devices.

Sharing the Images & Videos

When you are ready to share the images, simply tap iCloud in your share sheet and invite up to one hundred people to share in their enjoyment. Simply click a contact to add those you wish to see the images – or who want to upload theirs to the stream as well. Each new image will be broadcasted to the entire group. You and the invitees can view the images, videos and comments made by others swiping through the uploaded images effortlessly.

With people view you are able to manage notification settings, invitations and subscribers and contributors to the stream, so you know who is viewing the uploaded images in the stream. The activity view will give you access to the newest images, comments and updates in real time because you will be the first person alerted to new activity.

So how does iCloud Photo Sharing and Photo Streams fit into your new Apple TV purchase? Glad you asked. By accessing the Photo Stream from your Apple TV, you can view the group's photos, videos, comments and activity from your HDTV by broadcasting the content from your device through your Apple TV, so the entire room can enjoy your vacation, reunion or project photos and videos without crowding around your phone or tablet.

Enjoy the action, the fun and entertainment from your television set, instead of your device, so everyone – including those who were there, and those who were not – can enjoy the images in high definition and in a larger format, so the excitement pours out from your television screen.

Keep in mind that all of the images in your iCloud Photos can be streamed through your Apple TV, so even the most pedestrians images – like those of you eating a cheeseburger as big as your head – can be viewed on the big screen. What's more is that videos can be streamed to your television in the same way, so everything you collect from your iPhone, iPad or iPod touch can be displayed effortlessly onto your HDTV. This means you are able to watch all of your stunts, road trips and entertainment videos on your screen – while the room takes it all in from the comfort of your couch, instead of crowded around your device.

Deleting Photos and Photostreams

You may decide you want delete individual photos or photostreams right on your Apple TV and from your iCloud account. To do so, simply highlight the photo or photo stream. Hold down the select (round) button on your Apple TV remote. A pop-up menu will come up with several choices including "Delete Photo" or "Delete Album." Press on the select button on your remote to select the choice. (Note: Any photo or photo stream you delete on your Apple TV will also be deleted from your iCloud account and any other devices that are synched with it.)

Apple TV Tips and Tricks

As with any electronic device, Apple TV comes with its own set of tips and tricks to keep your usability and functionality operating optimally. Here are a few that will help along the way.

How to Set Up and Use Bluetooth

Bluetooth technology is a lot of fun, and pairing it with your Apple TV will allow you to enjoy the very ends of your entertainment means. All you have to do is sync your Bluetooth device.

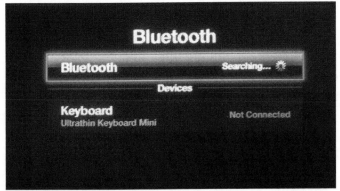

The following example relates to a keyboard, which can be helpful in terms of entering text (when required) onto your Apple TV's various input screens. Here's how to pair the keyboard:

1. Turn on your keyboard and if necessary ensure that it is discoverable. A blinking LED indicates that the Apple Wireless Keyboard is in the discoverable mode.
2. On your Apple TV, select Settings > General > Bluetooth.

3. Select Keyboard (or the device you are pairing) from the list.
4. Enter the four digit pairing code required.
5. Watch as the item connects to Apple TV.

How to Update System

Updating your Apple TV operating system is incredibly important to your viewing pleasure, as it ensures that programming and functionality are completely up to date for your enjoyment. Here is how:

1. Select Settings > General > Update Software.
2. Apple TV checks for an available update, if one is available. If so, a download message should appear.
3. Click Download Now to start the download process.
4. Never disconnect your Apple TV during the updating process.

Move Channels for Organization

You don't have to keep the channel layout on your Apple TV the way it is showing up on your screen. In fact, you can rearrange the channels to match your preferences. For example, you can have the channels you watch most towards the top, and the ones you watch least towards the bottom, or you could arrange the channels in alphabetical order.
To move a channel:

1. With your Apple TV remote, hover the selection box over the channel you'd like to move.

2. Press and hold down on the solid gray circular button in the middle of the directional arrows on your Apple TV remote. The selected channel will begin to shake or bounce.
3. Use the directional arrows to move the channel left, right, up and down, until you have it in the desired spot.
4. Press on the solid gray circular button to place the channel there. It will now stop shaking and be in its new spot.

Repeat this process with each channel you want to move on your screen.

How to Use the Fun Stuff!

Apple TV is more than just programming! You can use AirPlay to bounce items from your tablet or phone to the big screen, making your entire iOS suite of devices completely functional with your newest Apple family member. Here are some of the best options available with Apple TV!

How to Use AirPlay

AirPlay allows you to stream photos, videos and music to your Apple TV wirelessly, so you can share all of your content with the entire room – instead of everyone hovering around your phone or tablet.

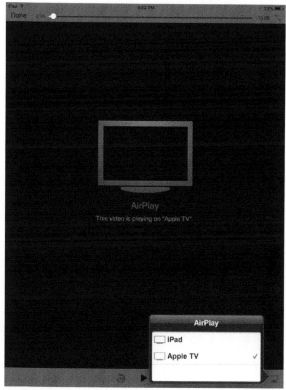

1. Turn your Apple TV On
2. Swipe the Control Panel Up on your iPhone or iPad
3. Select the AirPlay Icon from the Lower-right
4. Tap the Apple TV Icon
5. What you Show on Your Personal Screen will be Mirrored on the Television Screen

Note: If you are playing a video clip or movie from your iOS device, the AirPlay icon will appear in the lower-right hand corner, so all you have to do is tap it with your Apple TV on to mirror the content.

How to Use Amazon Instant Video

Apple TV doesn't have the Amazon Instant Video channel as part of its standard lineup of channels, but that doesn't mean you can't enjoy it on Apple TV with this tip. If you've got an iPad, iPod Touch, iPhone or Mac computer with the Airplay feature, you can still watch Amazon Videos on your Apple TV.

1. Ensure you have Airplay set up in the Apple TV settings and that you can access it with your particular iOS device or Mac computer/laptop.
   *For MAC laptops and computers with the Mavericks OS, you can go to System Preferences > Displays and set the Airplay to On or Off, as well as set it to "show mirroring options in the display bar when available."
2. Go to the App Store on your mobile device to get the Amazon Instant Video app (on a computer you can simply go to the Amazon website on a web browser and log in there).

3. Open Amazon Instant Video on your mobile device, or on your web browser, and log in with your account info. You will be presented with your video library as well as options to rent or buy movies and TV shows. If you're an Amazon Prime member you'll also have the ability to watch free movies streaming online.
4. Begin playing a movie or TV show video on your particular device. ("Watch Now" button)
5. Look for the Airplay symbol on your device screen (usually in the lower right-hand corner). Tap on it and choose "Apple TV." The Amazon video should now be playing on your Apple TV to your TV screen.

Note: You can also pause/play, select closed captions (when available) and use other on-screen options to enhance the Amazon video viewing experience from your device.

How to Use Remote App

You can turn your mobile device into a touchscreen remote to control your Apple TV, as well as play iTunes Radio or iTunes music/videos from anywhere in your home. There is a special app called "Remote" by Apple on the App store which will allow you to do all of these things. The App is free to download and install on your particular mobile device including the iPhone 5 or 5S and the latest iPads or iPod Touch devices.

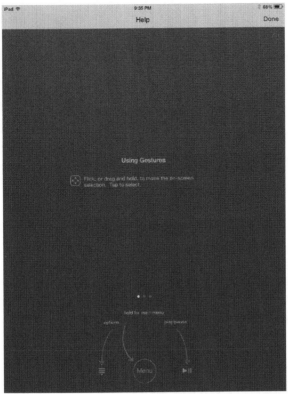

Once installed, open the App and you may be presented with various on-screen instructions about how to control your Apple TV (see image above). You will also see your library of available iTunes music which you can choose from to play on your Apple TV. Tap on any particular album or song cover art, title, musician or playlist to begin playing the music of your choice.

In the upper right hand corner of this app there's also a four-way directional arrow. Tap on this to bring up a touchscreen remote which is similar to your Apple TV remote.

With the touchscreen remote, dragging your finger on the screen in the desired direction will move you around on the Apple TV screen in that direction. Tapping on the screen will make a selection, for example if you were trying to select a channel or particular content to watch.

On the bottom of the remote screen are three icons. The first to the very left is for any available options you might have on that Apple TV screen. You can also tap on the "Menu" button to get to your main menu, or the previous screen, or the play/pause button to control any content your playing.

Using iPhone/iPad/iPod with Apple TV

Using personal device works hand and hand with Apple TV, as demonstrated in the AirPlay and Remote App examples. They also allow you, through the Home Share and iCloud technology, to access all of your favorite programming, images and videos from any iOS device you have – on the go, or from your living room! Take advantage of each outlet, and keep all of your operating systems up to date, so you will always have access to what is important – or at the very least, entertaining – in your life.

How to Reset and/or Restart Apple TV

There may come a time when you have to reset your Apple TV. Never fear, as electronic devices do not always play nicely, and in this case it is easy to do.
Restart your Apple TV:
1. Select Settings.
2. Select General.
3. Select Restart.

Allow the device to reboot completely, without unplugging it. If you would like to reset the device completely, and give it to a friend or sell it without all of your information remaining behind:
1. Choose Settings.

2. Choose General.
3. Choose Reset and then click Restore.

Your Apple TV will restore to factory settings and will download and install the latest Apple TV software. Do not disconnect your Apple TV's power cable or turn the system off during the restore process.

Jailbreaking the Apple TV

The term "jailbreaking" refers to a way in which users can hack the Apple TV and have greater control over the device. For the most part, it helps hack into the Apple TV's iOS (operating system). For the 5.3 version of Apple TV's OS, it allowed users to run XBMC and ATV Flash. Jailbreaking may also unlock more features such as the ability to use extra apps and to tweak aspects of the overall Apple TV experience. While there are plenty of guides for how to do this online, it's generally not recommended that you jailbreak your device. First and foremost, it could violate any warranties you have on the device. Secondly, jailbreaking could ruin your device and require repair or replacement, costing more money. If you still feel you want to jailbreak your TV, it's best to find a trusted online source and/or have someone who has experience with technology around to help out with the process.

See more information about Jailbreaking Apple TV at idownloadblog.com website, including a YouTube video walking users through the process.

Trouble Shooting Apple TV

Again, technology is not perfect. Sometimes trouble raises its ugly head. Instead of becoming frustrated, use these easy steps to getting your picture and entertainment back on track. Here are the basics, before you try anything else!
First, make sure:

- The audio and video cables between Apple TV and your TV are pushed in all the way.
- The power cords for Apple TV and your TV are securely connected to a working power source.
- Your TV is turned on and switched to the correct input.
- Apple TV is connected to your network. Go to the Settings menu on Apple TV, select Network, and see if Apple TV has an IP address.
- Your network and Internet connections are on and working properly.

TV Screen Fuzzy or Black

If your television screen appears fuzzy or black, you may need to select a video mode your TV supports.
To select a video mode:

- Press and hold both Menu and + (Plus Sign) on the Apple Remote for approximately six seconds
- Press + (Plus Sign) or – (Minus Sign) on the Apple Remote to Process the display resolutions
- When Apple TV reaches an acceptable display resolution and "If you can see the Apple logo, select

OK" appears on your TV screen, press the play/pause button

Apple TV Not Responding

If you can see a picture, but your Apple TV is not responding, try pressing and holding Menu Button on the Apple Remote to return to the Apple TV main menu.

Video with No Sound

If you are not hearing sound, but content is playing, the source of the problem could be a number of things, and the following checklist will help you determine which one applies to you.

- If your Apple TV is connected to an A/V receiver, make sure the receiver is turned on.
- Ensure the input setting you have selected on your receiver matches the input you have your audio cables connected to.
- Ensure sure the volume on your television or receiver is turned up and is not inadvertently muted.
- Ensure you are using the correct audio cable, and that it is connected firmly to Apple TV and to your television.
- If you are using the HDMI port on your television and Apple TV, ensure your TV supports audio through its HDMI port. The HDMI ports on some older televisions only support video.

- Ensure that you are using the proper remote. The same remote that serves as the "input" button to change your viewing to Apple TV is the one that will control the volume – not your cable or satellite remote.

Remote Not Working

- If you have paired Apple TV and an Apple Remote, make sure you are using the paired remote.
- If you are, indeed, using the paired remote and the Apple TV status light flashes white when you press buttons on the remote, the problem is not with the remote. You may need to restart your Apple TV instead.
- If you are using an unpaired remote, the Apple TV status light flashes amber. Use the correct remote.
- Make sure the IR receiver on the front of Apple TV is not blocked, and is receiving the signal properly.

Network Issues

If you are having trouble accessing your network, it could any number of small problems. Here is the best place to start.

- Check the IP address Apple TV is using to ensure it matches your IP address. Sometimes during setup an extra digit can be added, or one can be inadvertently erased.
- Check for any obstructions and adjust the location of the base station or Apple TV.

- If security is enabled on the network, temporarily disable it on the base station and try connecting again.

Apple TV Accessories

Once you familiarize yourself with Apple TV, you will begin to understand how it fits your individual or household needs. Everyone uses the device differently, and at only $99 for the TV itself it is the perfect addition to an Apple-based household, not to mention one that wants access to their Hulu or Netflix accounts. These services are an exceptional way to enjoy programming on your television, instead of on your tablet individually and makes for great binge watching!

Although channels and apps do not directly qualify as Apple TV accessories, but more of the staples of its existence, you can find a few that fit your personal make-up and enjoy them as you would any great accessory!

However, sometimes there is no substitute for the real thing. If you are looking for something to enhance your system or viewing pleasure, the following accessories will fit the bill effortlessly.

Speaker Systems

Sure, your television may be putting out enough sound as it is, but what if you want to enjoy a surround sound experience while playing music? You can link your Apple TV via Bluetooth to a wireless speaker system of your choice, and place the speaker outside during parties, so all of your guests are listening to the same music – at the same time – no matter where they are on the property. It is a very savvy way to keep spirits up while entertaining, and you will not have to change the music at any time as they will be playing directly from your iTunes account. Pricing will vary based on the speakers you choose, their capabilities and their brand. One of the more popular choices is this JBL Flip Bluetooth speaker that sells for about $99 or less.

H-Squared TV Platform

This unique project allows you to mount your Apple TV anywhere, and aim it to any platform, so your signal is direct and accurate. It also works flawlessly in multi-media rooms, so the device is out of reach and unobtrusive. You can face the platform any way you see fit, which is perfect for larger rooms. It also provides easy access to ports and cables, so you are never fighting with a connection. The price is generally 29.99 or less. View H-Squared TV platform for more details.

Apple TV Display Stand

You may prefer to have your Apple TV set up in a unique way based on your home entertainment set up. Instead of placing it flat on a surface or using a mount kit to attach it to your television, you can use an Apple TV display stand to hold the device upright on a flat surface. This may also make it easier to connect wires to the back of the device for any future fixes or upgrades. See the Apple TV stands here for an idea of what these might look like.

HDMI & Sound Cables

HDMI cables will help you enjoy the hi-definition viewing experience with your Apple TV. Since the Apple TV doesn't include them, you may need to buy one to go with your particular television set-up. Make sure to find good quality HDMI cables that will be long enough to connect from your television to the area you place the Apple TV unit in. You can find quality HDMI cables at Amazon.com and other retailers for around $9.99 or so.

Additionally, you may have a stereo receiver or other compatible way to enjoy "optical audio." The back of your Apple TV may or may not have this output which you can get an optical audio cable for and plug into a stereo or other capable device. This could very well kick your Apple TV's sound up another notch.

Ethernet & Micro USB Cables

If you don't have a wireless internet network available in your household, you will need to connect your Apple TV directly to your cable or DSL modem. To do so, you'll want to have a spare Ethernet cable available. You may also want one for future troubleshooting if your wireless network doesn't seem to be connecting with your Apple TV.

Also, there is a Micro USB output on the back of your Apple TV. You can use a Micro USB cable to plug your Apple TV into your computer for further troubleshooting. These cables are generally inexpensive to purchase.

Apple TV Remote

You may need to buy an Apple TV remote, should you lose the one that came with your device, or if say you acquired your Apple TV used. An Apple TV remote can be purchased through auction websites such as eBay, or on Amazon.com for under $20. Remember that you will need to follow instructions to pair your new remote with your Apple TV, outlined earlier in this guide.

Believe it or not, there are even protective sleeves available to make sure you don't get any scratches or smudges on your remote!

AppleCare Protection

Purchase an extended two year warranty on service and support for only $29.

Apple TV has a lot to offer, and with new channels being added with nearly each update, the selection continues to increase. It also makes a perfect gift for the Apple lover in your life, who wants to expand their entertainment reach from beyond their laptop or tablet!

With a price starting at $99, the Apple TV creates a perfect outlet for home movies, videos and images to scroll during family gatherings, while music plays effortlessly at the same time. The more you play with your Apple TV, the more you will love everything it can do and add to your life. Grab one online today, or stop into local Apple retailer for a preview of the exciting things Apple TV has to offer. You will not be sorry!

Conclusion

Apple TV is a fun way to experience all of your favorite television shows through any of the included applications like Hulu Plus or Netflix, while streaming newly released movies and old favorites through purchase or even temporary rental status. Apple TV also gives you access to the information, photos and videos that you collect from your devices, allowing you to view the content on a larger format for exceptional enjoyment.

This small device is available for a mere $99 or less, and like most Apple devices is easy to set up, thanks to its intuitive software. All you have to do is plug in the device, attach it to your television, turn it on and enter your existing Apple account information.

The applications, including television, music and videos will keep you – and your guests – entertained for hours, prior to tapping into your personal content to share with those who will appreciate it most.

Whether you use your Apple TV for personal enjoyment, or to share content with your entire family, the device is set up just like your other Apple electronics, which means you can password protect its use, downloads and purchases to keep your family from tapping into sensitive – or costly – content.

Everyone uses the device differently, and with Apple's exceptional design, is able to personalize their Apple TV to suit their needs. No matter what you use yours for, there is plenty of versatility, variety, channels and application availability to suit the needs of the masses and to keep you and your family engaged as you see fit. The future looks bright for the streaming media device with news that developers will be able to add even more channels to Apple's lineup. Each Apple TV update brings new and innovative features for improved ease of use and entertainment. Enjoy the power of the tiny device that will fit anywhere, without getting in the way of your home's design or décor. You will be amazed at its power, and embracing it will open your entertainment value up to a whole new world.

More Books by Shelby Johnson

iPad Mini User's Guide: Simple Tips and Tricks to Unleash the Power of your Tablet!

iPhone 5 (5C & 5S) User's Manual: Tips and Tricks to Unleash the Power of Your Smartphone! (includes iOS 7)

Kindle Fire HDX & HD User's Guide Book: Unleash the Power of Your Tablet!

Facebook for Beginners: Navigating the Social Network

Kindle Paperwhite User's Manual: Guide to Enjoying your E-reader!

How to Get Rid of Cable TV & Save Money: Watch Digital TV & Live Stream Online Media

Chromecast User Manual: Guide to Stream to Your TV (w/Extra Tips & Tricks!)

Google Nexus 7 User's Manual: Tablet Guide Book with Tips & Tricks!

Roku User Manual Guide: Private Channels List, Tips & Tricks

11088619R00036

Printed in Great Britain
by Amazon.co.uk, Ltd.,
Marston Gate.